SIMULATING GRAVITY WITH THE ATWOOD MACHINE

The Experiment that Proves Galileo Wrong

BY

ELIPHAS PHIRI

MttC Publications

DEDICATION

To myself

Table of Contents

PREFACE

The text book Physics for Scientists and Engineers by R. A. Serway and J. W. Jewett opens its topic on free fall with these words; **it is well known that, in the absence of air resistance, all objects dropped near the Earth's surface fall toward the Earth with the same constant acceleration under the influence of the Earth's gravity. It was not until about 1600 that this conclusion was accepted. Before that time, the teachings of the great philosopher Aristotle (384–322 B.C.) had held that heavier objects fall faster than lighter ones.**

This is what we have been learning and teaching for 400 years. But Newton's Second Law, Newton's Law of Universal Gravity, the rules of vector addition and Atwood Machine show that Galileo was wrong and Aristotle was right.

Two types of Atwood machines can be used to represent gravity (or attraction) and antigravity (or repulsion) between two bodies. These machines can also be used to find the net acceleration or simulated g between two objects.

This led me to conclude that Galileo must have been wrong; objects do not fall at the same rate. They only seem to fall at the same rate. This book reveals the reasons why the objects seem to fall at the same rate.

There is a God who Reveals Mysteries [Hidden Things] Daniel 4

January 2024

EP

CHAPTER 1
The Relativity of Mutually Attractive/Repulsive Objects

The Net Force of Mutually Attracting Objects

For forces which are attracting, the forces add up. If the forces F_1 and F_2 are attracting each other. The net Force is $F_n = F_1 + F_2$ and not $F_1 - F_2$.

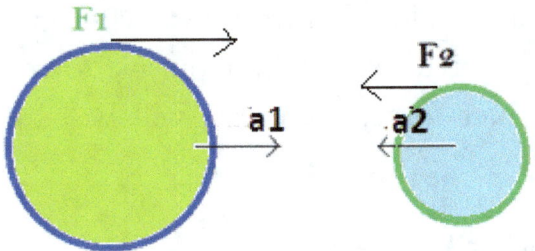

Suppose we have two objects attracting each other as shown, if the green object is kept still or fixed and the blue is left mobile. The force F_2 will have a reaction force on the blue object in the same direction as F_1 but equal to F_2. The resultant force, therefore, will be the sum of the two forces i.e.

$$F_n = F_1 + F_2$$

The Net Force of Mutually Repulsive Objects

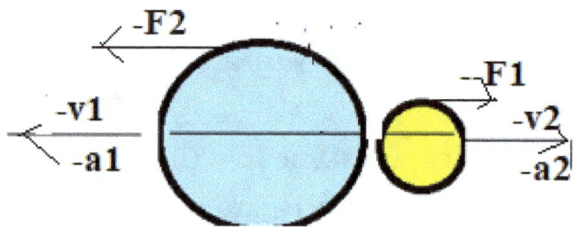

Similarly, for repulsive forces the net force will be the sum of the forces but negative in value. The direction of attractive forces can be taken as positive so that the repulsive force is negative. Net force, therefore, will be $-F_1 - F_2$,

$$F_n = - (F_1 + F_2)$$

Moderated Net Acceleration of Mutually Attracting Bodies

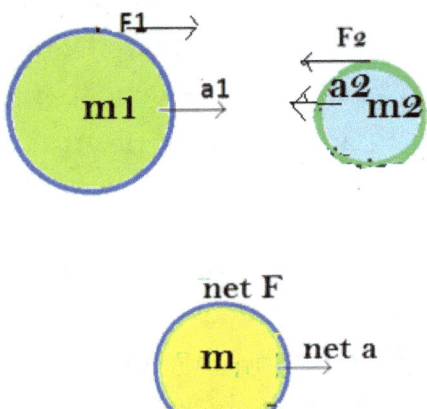

equivalent
mass

Suppose we have two objects attracting each other as shown. The effect of F_1 will be to make object m_2 move with acceleration a_1, and the effect of F_2 will be to make the object m_1 move too with acceleration a_2. These movements will be in the opposite direction of each other. The accelerations a_1 and a_2 will be determined by the amount of inertia or masses of the two objects.

The movements of the two masses can be

simplified into one movement, a net movement. We can consider the whole movement as that of one object with net mass, m_n, of the sum of the two masses i.e. $m_n = m_1 + m_2$. This implies that the net force, F_n, on m_n will be equivalent to the sum of the forces F_1 and F_2 acting on m_n i.e. $F_n = F_1 + F_2$.

The resultant acceleration, a_r of m_n, however, will be affected by the masses of the two objects. It will not simply be the sum of the two accelerations of the two objects i.e.

$$a_r = a_n = a_1 + a_2$$

as I previously stated. Instead the net acceleration will be moderated by the masses. How it is affected by the masses has been mind boggling until now. The following derivation shows how the net mass affects the net acceleration; $F_n = F_1 + F_2$

$F_n = m_1a_1 + m_2a_2$, but $m_1a_1 + m_2a_2$ is not equal to $(m_1 + m_2)(a_2 + a_1) = m_1a_1 + m_2a_2 + m_1a_2 + m_2a_1$ because it does not mathematically stand. So the resultant acceleration a_r cannot be equal to $a_1 + a_2$.

The resultant acceleration, a_r, will also be less than $a_1 + a_2$ since $(m_1 + m_2)(a_2 + a_1)$ is bigger than or equal to $(m_1 + m_2)a_r$ at the least i.e.

$$(m_1 + m_2)a_r \leq m_1a_1 + m_2a_2 + m_1a_2 + m_2a_1$$

Moreover, by the conservation of energy, we would not expect the net acceleration to be greater than its components unless extra energy is applied to the system. So net acceleration cannot be greater than $a_1 + a_2$. It can be equal to $a_1 + a_2$ at the most. So the resultant acceleration a_r is;

$$a_r \leq a_n \quad \text{or} \quad a_r \leq a_1 + a_2$$

We saw that m_n is the equivalent mass of the two masses in the system. So $m_n = m_1 + m_2$ and $F_n = F_1 + F_2$. Now there must be a certain acceleration, a_r, such that $F_n = m_na_r$, which gives $m_na_r = (m_1 + m_2)a_r$, but $F_n = F_1 + F_2$ also, which is $F_n = m_1a_1 + m_2a_2$, $m_na_r = m_1a_1 + m_2a_2$, then equating $(m_1 + m_2)a_r = m_1a_1 + m_2a_2$

And hence $\quad a_r = \dfrac{m_1a_1 + m_2a_2}{m_1 + m_2}$

So this is the net acceleration moderated by

the masses. We can therefore call it the moderated net acceleration a_m. It is not equal to the net acceleration i.e. $a_n \neq a_m$.

So $a_r = \dfrac{m_1 a_1 + m_2 a_2}{m_1 + m_2}$ becomes

$$a_m = \dfrac{m_1 a_1 + m_2 a_2}{m_1 + m_2}$$

Moderated Net Acceleration of Mutually Repulsing Bodies

For repelling objects, the repulsive forces will set up reaction forces R1 and R2 on the opposite objects. The net force will cause accelerations $-a_1$ and $-a_2$ on the two objects.

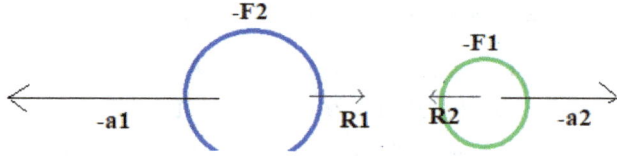

The net acceleration will therefore be;

$$a_m = \dfrac{-(m_1 a_1 + m_2 a_2)}{m_1 + m_2}$$

This is the moderated net acceleration that the object with the smallest force in the system will move.

Moderated Net Velocity of Mutually Attracting Bodies

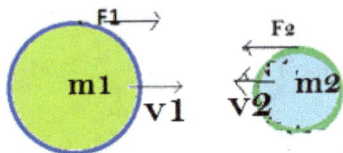

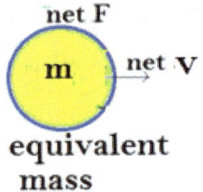

equivalent
mass

Suppose we have two objects attracting each other as shown. The effect of F_1 will be to make object m_2 move with Velocity v_1, and the effect of F_2 will be to make the object m_1 move too with velocity v_2. These movements will be in the opposite direction of each other.

From $a = \dfrac{m_1 a_1 + m_2 a_2}{m_1 + m_2}$ we can replace a with

v/t to get $\dfrac{V_m}{t_m} = \dfrac{m_1\dfrac{v_1}{t_1} + m_2\dfrac{v_2}{t_2}}{m_1 + m_2}$, $V_m=$

$$\frac{m_1\dfrac{v_1}{t_1} + m_2\dfrac{v_2}{t_2}}{t_m(m_1 + m_2)}$$

Moderated Net Velocity of Mutually Repulsing Bodies

For repelling objects, the repulsive forces will set up reaction forces R1 and R2 on the opposite objects. The net force will cause velocity $-v_1$ and $-v_2$ on the two objects.

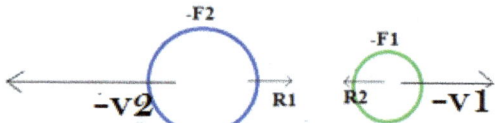

The net velocity will therefore be;

$$V_m= \left| -\frac{m_1\dfrac{v_1}{t_1} + m_2\dfrac{v_2}{t_2}}{t_m(m_1 + m_2)} \right.$$

Net Distance of Mutually Attracting Bodies

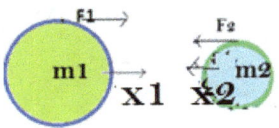

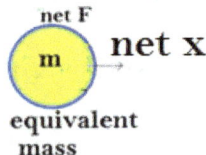

equivalent
mass

Suppose we have two objects attracting each other as shown. The effect of F_1 will be to make object m_2 move a distance of x_1, and the effect of F_2 will be to make the object m_1 move a distance of x_2. These movements will be in the opposite direction of each other. The net distance will be;

$$x_n = x_1 + x_2$$

Net Distance of Mutually Repulsing Bodies

For repelling objects, the repulsive forces will set up reaction forces R1 and R2 on the opposite objects. The net force will cause distances $-x_1$ and $-x_2$.

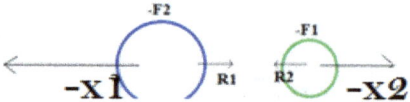

The net distance will therefore be;

$$x_n = -(x_1 + x_2)$$

Net K.E. of Mutually Attracting Bodies

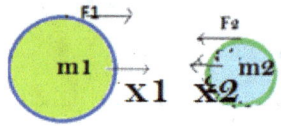

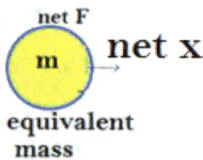

equivalent
mass

F_1 will make object m_2 move with velocity of v_1 and hence KE of $\frac{1}{2}mv_1^2$. F_2 makes the object m_1 move with a velocity of v_2 and hence KE of $\frac{1}{2}mv_2{}^2$. The net K.E. will be;

$$\boxed{KE_n = KE_1 + KE_2}$$

$$\frac{1}{2}m_1v_{1_2} + \frac{1}{2}m_2v_{2_2} = \frac{1}{2}m_nv_{n_2}$$

$$\frac{1}{2}(m_1v_{1_2} + m_2v_{22}) = \frac{1}{2}m_nv_{n_2}$$

$$KE = \frac{1}{2}(m_1v_{1_2} + m_2v_{2_2})$$

From the above formulae we can get net velocity as;

$$m_nv_n 2 = (m_1v_{12} + m_2v_{22})$$

$$v_n{}^2 = \sqrt{\frac{(m_1v_1^2 + m_2v_1^2)}{m_n}}$$

$$v_n = \sqrt{\frac{(m_1v_1^2 + m_2v_1^2)}{m_2 + m_1}}$$

KE of Mutually Repulsing Bodies

For repelling objects, the repulsive forces will set up reaction forces R1 and R2 on the opposite objects. The objects will move distances $-x_1$ and $-x_2$ respectively.

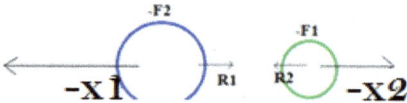

The net distance will therefore be;

$$x_n = -(x_1 + x_2)$$

CHAPTER 2

Galileo was Wrong: Masses do not Fall at the Same Rate

Net Force due to Gravity, F_n

Newton said that all objects attract each other i.e. all objects have gravity. When placed near each other, the objects will therefore fall into each other. If object 1 has force 1 and object 2 has gravity force 2 then the net force will be $F_1 + F_2$. The net force therefore between the two is greater than their individual gravitational forces.

So
$$F_n = F_1 + F_2$$

This net force will be different for two objects of different masses brought near the same object because of differences in their forces of gravity.

So the object-earth system with greater net force will have higher rate of falling than the object-earth system with less net force.

The Complete Definition of Weight

The force the earth exerts on the object near the surface is called the weight. So $F_1 = W_1$ and the F_2 can be called the weight of the earth on 'planet object' so that $F_2 = W_2$. So

$$F_n = W_1 + W_2$$

But $W = mg$ so we get;

$$F_n = m_1 g_2 + m_2 g_1$$

or $$F_n = m_e g_o + m_o g_e$$

for objects near each other.

But the force between two objects is also given by $F_g = G\dfrac{M_1 M_2}{r^2}$

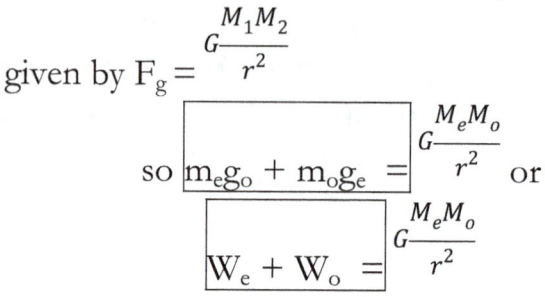

so $\boxed{m_e g_o + m_o g_e} = G\dfrac{M_e M_o}{r^2}$ or

$\boxed{W_e + W_o} = G\dfrac{M_e M_o}{r^2}$

where W_e and W_o can be called the particular weights of the earth and object respectively. The definition of weight is given as; $M_o g_e$ and is equated to $G\dfrac{M_e M_o}{r^2}$. But this definition ignores force of the object on the earth. The complete definition of weight, therefore, should include the force of the object on the earth or planet. The complete definition should be the sum of the two forces involved in the 'tug-of-war'. It should

be; $\quad m_e g_o + m_o g_e = G\dfrac{M_e M_o}{r^2}$

i.e. $\quad \boxed{W = m_e g_o + m_o g_e}$

OR

$\boxed{W = W_e + W_o}$

We must be talking about the net weight therefore and not just weight. All weights we measure are net weights. The paradigm must shift.

The Weights of the Planets

When we measure weight using a spring balance we are actually measuring this net force or net weight. The particular weights of the object alone or earth alone can never be found i.e. W_e and W_o. We can only approximate W_o of astronomically small objects since their masses are very tiny in comparison to the earth's mass. Therefore, the object is accelerated more than the earth or planet. This gives rise to a bigger weight for the object than the earth or planet. So the planet's particular weight on the object can be ignored.

The particular weight of the planet or earth on the astronomically small object is very small almost zero since the pull of gravity of the object acting on the planet or earth is very small and hence makes the planet or earth move very slowly or not at all. Thus the small acceleration gives rise to a very small weight. This weight cannot, therefore, be approximated. Moreover, it is not of practical significance to the scientist in relation to the weight of astronomically small or everyday objects. It seems the concept of weight is practically useless compared to the concept of mass. More comparisons can be done using mass than weight. Mass is more practical than weight.

When we measure the weight of the object we are simultaneously finding the weight of the earth/planet on that object. It is like the faces of a coin or the poles of a magnet; they can never be separated to stand on their own.

In short, weight is reciprocal. So the same net weight of the object on the earth/planet is also the net weight of the earth/planet on that object i.e. $m_e g_o + m_o g_e = W$. It is not the weight of the Earth/planet only or weight of the object only but of both.

For instance, for an isolated system of an object of mass 60kg and earth then the weight of the object on earth is approximately 60 x 10 = 600N and the weight of the earth on the object is also 600N. This 600N is the weight of the object plus the weight of the earth.

The weight of the object alone is not 600N. It is less than 600N and the weight of the earth on the object is not 600N but less almost zero.

Since it depends on the net g, it is possible for the complete weight of the earth or any other astronomical body to be small or even zero. Since the pull of the object on the earth/planet is small, the earth/planet accelerates slightly or remains stationary. The force/weight, therefore, will also be small or zero in spite of the large mass of the earth/planet.

Approximating g of Small Objects

This implies that we can get an approximate of the g of the object since the mass of the earth is known. We can ignore the weight of the planet on the object and use the formula $W = mg$. So for a 60kg object whose weight is approximately 600 which is equal to approximately $m_e g_o$. Since the mass of the earth has factor 10^{24}kg, we get g_o in factor of 10^{-22} N/kg or 10^{-22}m/s^2. For a 20,000 tonnes ship i.e. 20,000,000kg the weight is approximately 200,000,000N . So the g will be approximately in the factor of 24^{-16}N/kg or 24^{-16}m/s^2. We can readily see that even for some of the largest manmade structures the g is still very small. The difference between the g of a 60kg object and that of a ship would be in a factor of 24^{-16}m/s^2. This difference would be hard to observe for the two objects.

Moderated Net Acceleration due to Gravity, g_n

In the previous editions I had stated that an object m_1 coming from outside the other object's field, will accelerate towards m_2 with acceleration of a_1, and m_2 will accelerate towards m_1 with acceleration of a_2 and the net g will be the sum of the two gs i.e. $g_n = a_1 + a_2$.

But it has come to my attention that this cannot be since the accelerations of each object will be affected by the masses.

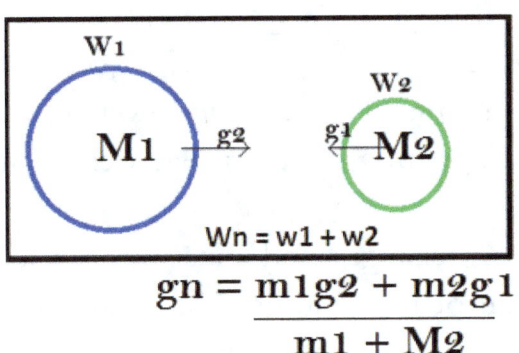

$$gn = \frac{m1g2 + m2g1}{m1 + M2}$$

If an object of mass m_1 with g of $\boldsymbol{g_1}$ is brought near the earth or planet with mass m_2 and g of g_2 then their masses merge together to give m i.e. $m_1 + m_2 = m$.

But their net g, g_n, will not be the sum of their gs i.e. $g_1 + g_2$ as we saw previously with mutually attracting objects. This is because the new weight of m, W_n, is not the sum of their weight, as we previously saw i.e. $W_1 + W_2 \neq W_n$ or $m_1g_2 + m_2g_1 \neq (m_1 + m_2)(g_2 + g_1)$. It does not mathematically stand. Instead, the net g will be a moderated g_m; so $(m_1 + m_2)g_m = m_1g_2 + m_2g_1$

$$g_m = \frac{m_og_e + m_eg_o}{m_o + m_e}$$

$$\text{or } g_m = \frac{1}{M}(W_o + W_e)$$

where $M = m_1 + m_2$ or mass of object and earth, W_o is $m_1 g_2$ or particular weight of object and W_e is $m_2 g_1$ or particular weight of the earth.

So the new g changes and must be smaller than the sum of their old gs. This is also the net g that an object-earth system will experience not $g_1 + g_2$ as I had previously stated.

This moderated net g experienced by an object when falling on earth or any other planet is also the g experienced by the earth or planet. We must no longer talk about a specific g of the earth or object. The only g we can talk about is the g net of each earth/planet-object system. The gs are specific to the masses and not general.

Galileo was Wrong

This clearly shows that Galileo was wrong; objects do not fall at the same rate. The g_m of the bigger object is bigger than the g_m of the smaller object. In short, heavier objects fall faster than less heavy objects.

But because we deal with mostly tiny masses (which I will call, astronomically small masses since they are considerably smaller than the earth no matter how big they seem to us), the difference between their masses are negligible and hence their net gs differ slightly as to give approximately the same net g when plugged into the formula; $g_m = \dfrac{m_o g_e + m_e g_o}{m_o + m_e}$.

Galileo Galilei was wrong. His famous experiment was an oversimplification which led to error because it was not accurate enough.

Heavier Objects Fall Faster (an Analytical Proof)

In defending Galileo's assertion that masses fall at the same rate in spite of differences in their masses, someone claimed that somehow, mysteriously, the masses and forces in gravity field balance each other in such a way that the acceleration due to gravity is always a constant. Someone else claimed that the bigger mass cancels out the expected higher acceleration by moderating the force since $\mathbf{g} = \mathbf{F/m}$ so that the g remains the same.

How that moderation or cancelling out is done has been mind boggling and a mystery until now.

This relationship $g_m = \dfrac{m_1 g_2 + m_2 g_1}{m_1 + m_2}$ shows how the moderation is done or how the g is affected. It proves Galileo was wrong and the mysterious moderation or cancelling out wrong. The moderation or cancelling is not in such a way as to make the g constant.

The relationship $g_m = \dfrac{m_1 g_2 + m_2 g_1}{m_1 + m_2}$

which we can rewrite as $g_m = \dfrac{m_o g_e + m_e g_o}{m_o + m_e}$ where m_o is mass of object, g_o is the g of the object , m_e is the mass of the earth and g_e is g of the earth, shows that moderated net g is not constant. It is moderated by the mass

through the relationship $g_m = \dfrac{m_o g_e + m_e g_o}{m_o + m_e}$.

All the objects we deal with are astronomically small and hence their g s are very small. When we replace this small mass and small g in the formula, the numerator in the formula $g_m = \dfrac{m_o g_e + m_e g_o}{m_o + m_e}$ gets smaller compared to the denominator. The denominator in the formula is minimally affected since we are adding a small mass to a huge mass. The numerator, however, is affected considerably and gets smaller since we are multiplying the mass or g of the earth with a very small number of g_o or m_o. The mass and g of the earth remains the same throughout but the objects have different masses and gs. So m_o and g_o varies but m_e and g_e does not.

It also seems that m_o is directly proportional to g_o so that as g_o tends to zero m_o also tends to zero though it lags behind since it is considerably larger than g_o.

So g_n tends to zero as g_o and m_o tends to zero i.e.

$$\lim_{m_o, g_o \to (0,0)} \frac{m_o g_e + m_e g_o}{m_o + m_e} = 0$$

But for a massive object like the moon or Jupiter, with more g, the numerator gets considerably bigger since we multiply the big mass or g of the earth by a big g_o or m_o. But since we are just adding the two masses on the denominator, the denominator will be smaller than the numerator. So the g_m gets big. It also seems that m_o is directly proportional to g_o so that as g_o tends to infinite m_o also tends to infinite.

So g_m tends to infinite as g_o and m_o tends to infinite i.e.

$$\lim_{m_o g_o \to (\infty, \infty)} \frac{m_o g_e + m_e g_o}{m_o + m_e} = \infty$$

Spreadsheet Simulation of *g*

The modulated acceleration formula can be used in spreadsheet to show the variation in net acceleration or simulated *g* due to a change in mass and corresponding change in *g* of that mass when it increases. The following table shows the changes in a simulated g as a result of changes in mass where the mass of the earth is considered as 24kg, for the sake of simplicity, and g of earth as 10N/kg and are kept constant. The values of the g of the object is taken to double as the mass doubles for simplicity 's sake.

Mo	g_0	Mod A
0.1	0.001	0.042
0.01	0.0001	0.0042
8	6	7
16	12	11.2
32	24	16
64	48	20.3

As the table shows as the mass and g of the object increases the modulated acceleration or g also increases. This shows that Galileo was wrong, heavier objects fall faster than less heavy ones.

The Flaws in Galileo's Experiment

Galileo's experiment on the Leaning Tower of Pisa had four flaws;

 i) The earth-object system was to be isolated for each object so that the gravity of one does not interfere with the gravity of the other.

ii) The difference between the two masses was not sufficiently great to show the difference in the rate of fallings. Because Galileo dropped very tiny masses with very small g in comparison to the earth, the net g for each dropping differed very slightly as to be imperceptible. If he had done the same experiment with a very tiny object like an atom and another massive object like the moon, the moon would have fallen faster than the atom i.e. g_m moon and earth system will be greater than g_m for an atom and earth system. The moon has g 1/6 th that of the earth and an atom must have g significantly smaller than that of the moon. So g net of moon-earth system will be greater than that of atom-earth system. In short, the moon will fall faster to the earth than the atom whose g net is less.

iii) The distance from which the objects were dropped was also not sufficiently great to show the difference in the rate of fallings. If two smaller objects smaller than the earth by a factor of a billion and hence their gravities were of a factor of let's say 10^{-6} but differed slightly (like Galileo's feather and stone or Scot's feather and hammer), then by; $y = -0.5gt^2$, they would fall a distance of ;

6×10^{-5}m in 120 sec i.e. 2 minutes, and

6×10^{-4}m in 1200 sec or 20 minutes, and

6 x 10⁻³m i.e. 0.006m or 0.6cm or 6mm in 200 minutes or 3.3 hours.

So Galileo would have required a very high height or very big difference in the masses of the objects to detect the difference in the falling rates between them. A height which would give a difference of 0.6cm or 0.006m would require a height of;

$$y = 0.5(9.81)(12000)^2 = 706,320,000m$$

or 706,320km.

So just to observe a difference of 6mm in the falling rates for small objects would have required a height or ramp more than 700 thousand km up into space, and a waiting time of more than 3 hours to see just one falling object reach the ground.

In short, there was a slight difference in the rate of falling of the feather and stone but Galileo could not easily observe it. With the speed at which the object hit the ground, Galileo could not have possibly differentiated the rates with his naked eyes. But to his credit in his rigorous experiment, he managed to slow the speed using ramps. But they were not sufficiently high and the objects did not sufficiently differ in mass. They worked correctly in determining that the rate is exponential in form but it is not the same for all objects. But one can claim that proves that g is approximately constant. But we have seen that if the difference in masses between the two objects being compared is great and the experiments done one at a time then their rate of fallings will not be constant.

So it's not the height of dropping which is a crucial factor but the difference in the masses being compared. The smaller the difference the less the difference in the rate of fallings and the more the difference the more apparent and great will be the difference of the rates of fallings of the two objects.

 iv)Galileo did not know Newton's law of Universal Gravitation. He did not know that the earth was also falling into the feather and the stone. He did not take into consideration the movement of the earth into the opposite direction.

Newton, too, could have proved that Galileo was wrong. But I guess he couldn't fathom challenging the great master. Newton could not see the implications of his own law of Universal Gravitational in relation to rate of falling bodies.

Contradictions if Galileo was Right

But someone may object. They may say that the mass or inertia of the moon or massive object may actually slow down the rate of falling hence cancelling out the effect of the extra mass and making the rates of falling of all objects equal. But in whose favor will the cancelling out be? Moon or earth? If we maintain the status quo and say all objects fall on a planet with a constant g of that planet then we would expect the earth to fall into moon with the g of moon which is less than that of the earth. The observer on the earth will therefore see the moon as falling into the earth slower than usual i.e. at the g of the moon $1.64m/s^2$. We cannot expect the moon to fall at the g of the earth because we are considering the earth as the object.

Or we can consider the moon as the object falling on the earth with the g of the earth since all objects fall on the earth with a constant g of 9.81m/s^2. We, therefore, have a contradiction and we are stuck in a quandary. So this cannot be. We cannot maintain the status quo. The moon cannot fall at g of the earth and at its own g at the same time. It must fall at a certain g which must be some combination of the two. Galileo was therefore wrong; objects do not fall at the same rate.

Moreover, we have already seen that the relationship $g_m = \dfrac{m_o g_e + m_e g_o}{m_o + m_e}$ shows that the heavier object will fall faster than the less heavy object on earth or a planet. So the moon will fall on earth at approximately;

$$g_m = \dfrac{m_o g_e + m_e g_o}{m_o + m_e},$$

$$g_m = \dfrac{10^{22}(10) + 10^{24}(1.64)}{10^{22} + 10^{24}}$$

We usually deal with astronomically small objects which are very close in g. So Galileo's stone and feather were relatively very close in mass even though to us they seem to have big difference in mass. The difference in masses was not big enough to give a perceptible difference in the falling rates.

Nevertheless, we can restate Galileo's assertion as; objects with relatively small difference in their masses and hence in their gs fall approximately at the same rate whether those objects are astronomical or microscopic in size.

If the difference in masses between them is considerably big then the rates of falling will also be considerably big and perceptible to human beings.

PROOF FROM ATWOOD MACHINE

The formula for Atwood machine of two masses connected together over a massless pulley agrees with this relationship.

For the Atwood machine with masses of different sizes as shown;

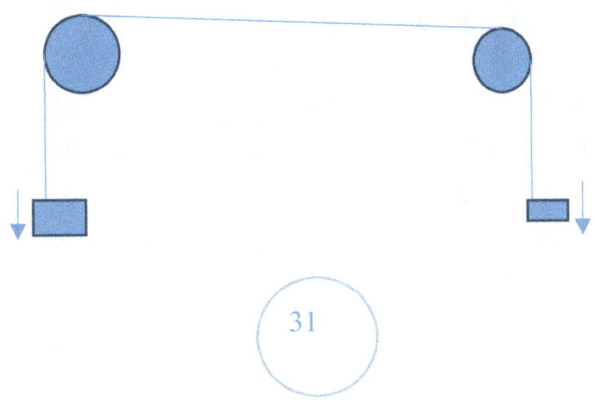

The net acceleration is given as;

$$a = \frac{m_1 g - m_2 g}{m_1 + m_2} \quad \text{for } m_1 > m_2.$$

This agrees with our formula except that the motions of the masses relative to their direction of the applied forces are both positive so they add up and the gs are of different sizes. The gs in our formula are both positive because the directions of motions are positive relative to the applied forces.

The Atwood formula also shows that the net acceleration of the falling object will not be constant. The net acceleration will depend on the difference between the two masses. If the two masses are of the same mass in magnitude ie if the difference in masses between them is zero the Atwood formula for our system clearly shows that the net acceleration is also zero.

If the net mass increases ie non-zero, then we would expect the net acceleration to increase either to the positive or negative side. The bigger the net mass the bigger the net acceleration. So a system of a very massive object and an extremely less massive object will have the most net acceleration.

This proves that g of the earth for specific objects cannot be constant. The earth-object system can be considered as an Atwood Machine in reverse.

The Reverse Atwood Machine [the Gravity (Attraction) Machine]

The Gravity Machine or reverse Atwood machine can be constructed as shown. Two pulleys with masses m_1, m_2 and m_3 can be used.

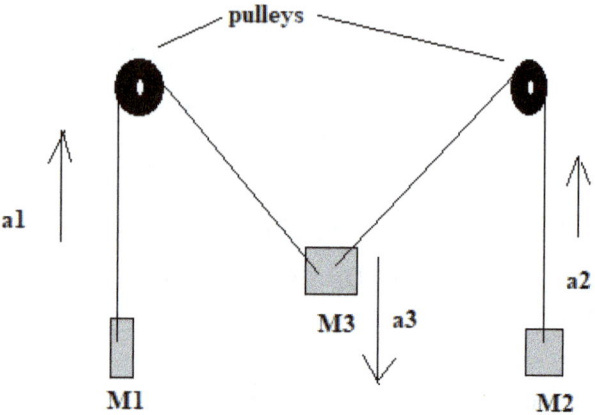

This machine is an analogue of gravity or attraction between two bodies and can be used to prove the modulated acceleration formula and hence simulate the modulated g between a planet and an object.

The force pulling m_1 and m_2 together is the force pulling m_3 down

$$\text{ie } F_3 = F_1 + F_2.$$

$$m_3g = m_1a_2 + m_2a_1 ,$$

In a real attractive case where there m_3 pulling the two masses together, a force, $(m_1 + m_2)a_m$, can be found which can be equivalent to m_3g

so that

$$(m_1 + m_2)a_m = m_1a_2 + m_2a_1$$

where a_m is the moderated acceleration. So

$$a_m = \frac{m_1a_2 + m_2a_1}{m_1 + m_2}$$

Anti-gravity, Anti-weight and Anti-g

From our section in chapter 3 on repulsion of bodies, if we have two repelling astronomical bodies then we can talk about anti-gravity. The force of repulsion could be called anti-gravity, the net acceleration between them could be called anti-g and the force between them could be called anti-weight.

Anti-Gravity [Repulsion] Machine

An Atwood machine can be made to show repulsion or simulate anti-gravity. Two cylinders can be connected in such a way as when one moves to the left it causes the other to move to the right hence simulating repulsion. A string connected to one cylinder can go over a bar or pulley wheel and connect to the second cylinder from the opposite side. The same can be done to the first cylinder also as the diagram below shows. The arrows show the direction of motion.

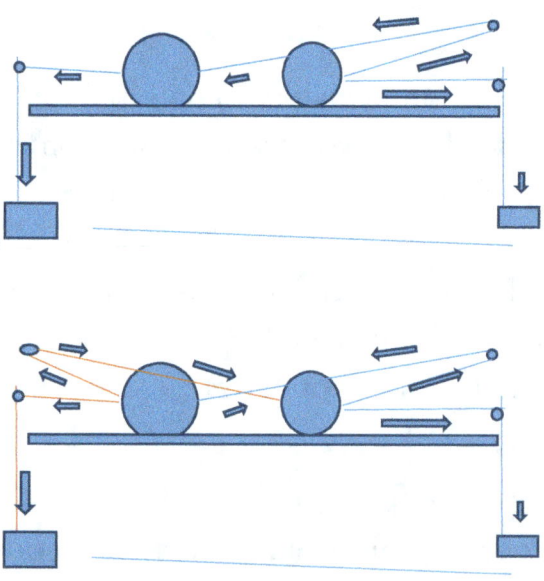

Gravitational Mass is not Equal to Inertial Mass

The relationship $g_m = \frac{1}{M}(W_o + W_e)$ becomes

$$g = \frac{W}{M} \text{ for earth}$$

W is the weight of the object we obtain by spring balance. The M is called the gravitational mass which is said to play the role of determining gravitational field strength. This gravitational mass has been said to be equal to the inertial mass of the object given by $\mathbf{F} = \mathbf{m_o a}$. But since we have seen that M is actually the sum of the mass of the earth and the mass of the object i.e.

$\mathbf{M = m_o + m_e}$, inertial mass is not equal to gravitational mass i.e. $\mathbf{M \neq m_o}$. Gravitational mass is extremely huge. The gravitational mass of astronomically small objects is slightly greater than the mass of the earth and extremely greater than the inertial mass of the objects. Now the name makes sense.

The Reciprocity of Gravitational Mass

Gravitational mass is also reciprocal just as weight i.e. the gravitational mass of the earth is also the gravitational mass of the object.

Factors that Determine Size of g_0

It seems that the accelerations due to gravity of objects and planets are directly proportional to the masses and density of the objects or planets. For instance g of earth is 9.81m/s^2, for the moon it is 1.64m/s^2.

$$\text{So} \quad g_0 = k\rho m_o \,, \quad g_0 = k\frac{m_o}{V} m_o$$

$$, g_0 = km_o^2/v, \text{ but } V = \frac{4}{3}\pi r^3,$$

$$\text{so } \boxed{g_0 = k\,3m_o^2/4\pi r^3}$$

k is constant of proportionality which can be called constant of g or constant of weight. M_o should not be confused with mass of the object for which the weight is being found in relation to that planet.

Weight can also be found on any object and not only on a planet. So the notation of g_o and m_o is more appropriate where O stands for object. All this implies that weight must be actually lower than what we measure it.

This formula shows that the magnitude of g in the object varies directly as the square of the mass and inversely as the cube of the radius from the center of the earth or object i.e.

$$g_o = k_w m_o^2 / r^3$$

k_w being a new constant of the weight.

Measurement of g

The measurement of g by pendulum is affected by this fact. The mass of the bob and it's gravity is tiny to show the real g of the system. More-over everything will be affected by the earth-universe system so the net g is the result of all the gravity of the universe and the earth.

Measurement of g and New Formula

The measurement of g by pendulum is affected by this fact. The mass of the bob and it's gravity is tiny to show the real g of the system. More-over everything will be affected by the earth-universe system so the net g is the result of all the gravity of the universe and the earth.

The period of the pendulum will be affected by the mass. It is no longer independent of the mass. If we replace $g_o = k\rho m_o$, into

$$T = 2\pi\sqrt{\frac{l}{g}}$$ we get $$T = 2\pi\sqrt{\frac{l}{k\rho m}}$$ but $\rho = \frac{M}{V}$, So $g = km^2/V$

$$T = 2\pi\sqrt{\frac{Vl}{KM^2}}$$ but for a spherical bob $V = 4/3$

πr^3, $$T = 2\pi\sqrt{\frac{4\pi r^3 l}{3KM^2}}$$, $$T = 2\sqrt{\frac{\pi(2\pi^2)r^3 l}{3KM^2}}$$, $T = 4$

$$\pi^3\sqrt{\frac{r^3 l}{3KM^2}}$$, $$T = \frac{4\pi^2}{9k^2}\sqrt{\frac{r^3 l}{M^2}}$$, $$T = \frac{4\pi^2\sqrt{lr^3}}{9k^2\ M}$$,

$$T = \frac{4\pi^2\sqrt{lr^3}}{9k^2\ M}$$, $$\left(\frac{2\pi}{3K}\right)^2$$ can be called the Pendulum Constant P.

$$\text{So } \boxed{T = P\frac{\sqrt{lr^3}}{M}}$$

This formula shows that the period varies inversely as the mass. We have been teaching that mass is not a factor, but when the mass is huge or when using a very sensitive pendulum, it is a factor.

The formula also shows that the period varies directly as the radius of the bob. This is just a consequence of the inverse square law of gravitational force.

Galileo's hunch, which he was actually testing

in the experiment at the tower of Pisa that the rate of falling is directly proportional to the density of the object was actually correct but the rate also depends on the mass. For the same substance with the same density the more massive object will fall faster than the less massive one. This is because the more massive object has more electrons and protons which cause more gravity force than for the less massive object.

Pendulum in a Vacuum

An experiment can also be done on earth using pendulums in a vacuum. Since the slowing down of a pendulum is due to air drag, a pendulum in a vacuum will swing forever or very long time until stopped by the small friction force between the particles in the string as it bends due to the swings.

Two pendulums can be constructed of the same length but different masses. The pendulums can be dropped from the same height. Since the bigger mass should drop faster, it will eventually overtake the smaller mass when it reaches the bottom. This overtaking may not be perceptible initially, but overtime the bigger g or acceleration of the bigger mass will make exponential increments in the difference of the two pendulums at the bottom. This will prove that the g is not constant and that the period is affected by the mass.

Rate of Falling Formula

From the equations of motion, the time of falling for objects into the earth will be;

$$t_o = = \frac{V_e\sqrt{g_e}}{g_e\sqrt{g_o}}$$

t_o, v_o being time for object and velocity of object. The time for earth falling into object will be;

$$t_e = = \frac{V_o\sqrt{g_o}}{g_o\sqrt{g_e}}$$

We do not have the sense of us moving into the opposite direction. To us only the moon or falling object will seem to be moving/falling.

For an observer on the earth or planet, objects therefore do not fall at the same rate.

It is like jumping up to catch a falling ball or object; the object is moving towards you at about 9.81 m/s^2 and you are also accelerating towards the object. You therefore meet the object earlier than if one did not jump. The net g of the jumping person and the falling object is therefore greater than 9.81m/s^2. The g of the ball to the person who did not jump was the normal g of 9.81m/s^2. To that observer objects fall at different rates. We have been wrong for more than 400 years.

Conservation of Position

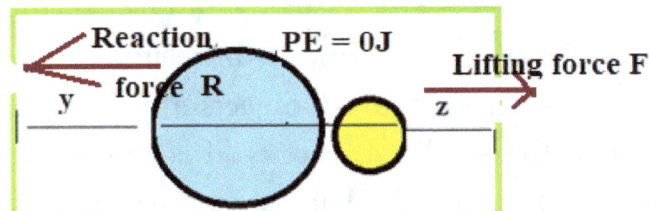

For objects in contact with each other and in the same field, when the object is lifted up from the earth, the earth also moves backward due to the reaction force by Newton's third law. When the objects falls back, the earth also moves towards the object due to the gravitational force. The two objects will meet again on the same position. So for an isolated system position of the objects is conserved.

The Reciprocity of Gravitational Potential Energy

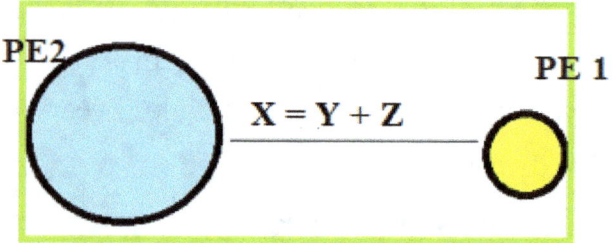

The PE of the two objects in relation to each other is zero when the objects are in contact. When the smaller object is lifted up, it gains potential energy PE1 in relation to the bigger object and distance Z , and $PE1 = M_og_ey$. But M_og_e is weight W_e.

The bigger object gets PE 2 in relation to smaller object and distance Y and $PE2 = M_eg_oy$. but M_eg_o is weight W_o, $PE1 = M_og_e$.

If one object is considered stationary and the other as moving then $PE = Mgy = Wy$. But

$$W = m_e g_o + m_o g_e,$$

$$\text{So } \boxed{PE = (m_e g_o + m_o g_e)y}$$

$$\boxed{PE = W_e y + W_o y}$$

$$\boxed{PE = Wy}$$

This PE is also the PE of the earth in relation to the earth i.e. PE is reciprocal just as weight.

When the objects fall back into each other, the PE is again back to zero. So the energy of the system is conserved and the position of the system is conserved i.e. the earth object system moves back to the original position.

Anti-gravity, Anti-weight and Anti-g

From our section in chapter 3 on repulsion of bodies, if we have two repelling astronomical bodies then we can talk about gravity. The force of repulsion could be called anti-gravity, the net acceleration between them could be called anti-g and the force between them could be called anti-weight.

The Anti-Gravity Potential Energy

If two objects are repulsing then we can talk about anti-gravity Potential energy. The formula will be the same as for the Gravitational PE except that it will be negative i.e.

$$APE = -(m_e g_o + m_o g_e)y$$

$$APE = -(W_e y + W_o y)$$

$$APE = -Wy$$

This PE is also the PE of the earth in relation to the earth i.e. PE is reciprocal just as weight.

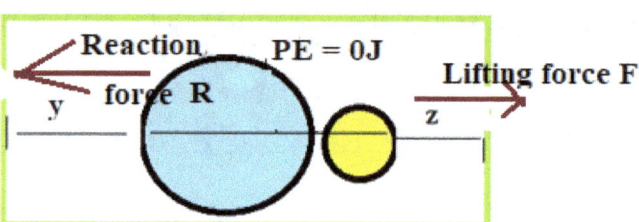

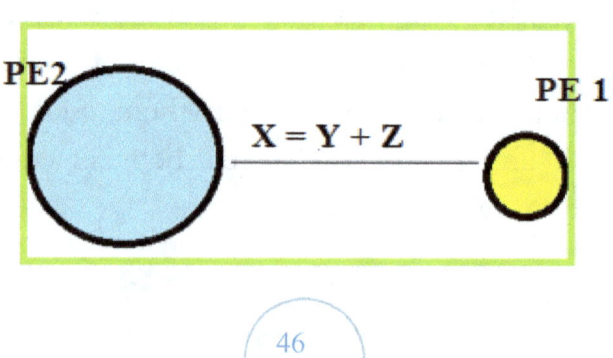

CHAPTER 3

RELATIVITY OF ACCELERATION

The Galilean Mutually Attracting Bodies

For two bodies in space each object is falling into another's gravity field at that object's g and the other object is also falling into that object with that object's g, they will meet faster and some way through the journey. The relative acceleration of one body from the observer on the opposite body, and vice versa, will be greater than if only one body has gravity and the other did not.

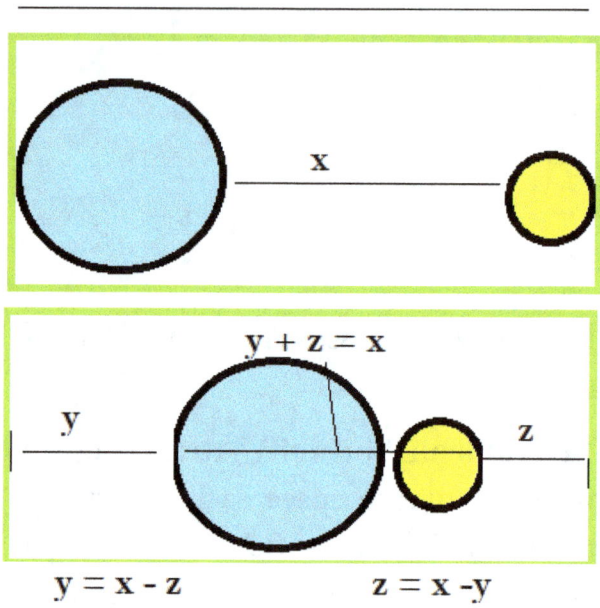

d

x

y + z = x

y

z

y = x - z   z = x -y

The blue object will fall into **z** by **y** meters while the yellow will move **z** meters. Instead of falling through the length x, the smaller object will seem to fall through $\mathbf{x} - \mathbf{y}$ to an observer on the yellow object, and instead of falling through **y**, the bigger object will seem to fall through $\mathbf{x} - \mathbf{z}$.

Also instead of taking the time t_x to meet, the objects will take t_{x-y} and t_{x-z} to meet. So instead of the acceleration being $a_{xe} = V_1/t_x$,

it will be $a_{x\text{-}z} = \dfrac{V_1}{t_{x\text{-}z}}$. Since V is the same but

$t_x > t_{x-z}$, $a_{x\text{-}z}$ is greater than a_{xe}. Also the acceleration of object a_{xo} will be different from $a_{x\text{-}y}$.

It is like jumping up to catch a falling ball or object; the object is moving towards you at about 9.81 m/s^2 and you are also accelerating towards the object. You therefore meet the object earlier than if one did not jump. The net g of the jumping person and the falling object is therefore greater than 9.81m/s^2 for an observer in space who does not see the ground as a reference point of the two motions. To that observer objects fall at different rates.

So for an observer on the yellow object, oblivious to his movement towards the blue object, will see the blue object fall faster than normal and vice-versa.

The Galilean Mutually Repelling Bodies
If the two objects are repelling then the relative velocity noticed by an observer on

one object at lower speeds will be the sum of the two velocities as they recede from each other i.e. $\mathbf{a} = \dfrac{-(v1+v2)}{t}$ instead of $\mathbf{a} = = \dfrac{-v1}{t}$. Since the time is the same in both cases but the observer notices one velocity as sum of the two velocities the net acceleration is bigger than expected.

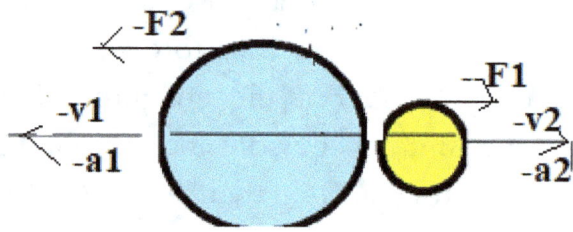

The Special-Relativity Acceleration Addition Formulae

For speeds near the speed of light the velocity addition of the repulsive objects reduce to that of the Lorentz Velocity Transformations equations; i.e.

$$U'_x = \frac{u - v}{1 - \left(\dfrac{v}{c^2}\right)u_x}$$

$$U_y' = \frac{u_y}{\gamma\left[1 - \left(\dfrac{v}{c^2}\right)u_x\right]}$$

$$U_z' = \frac{u_z}{\gamma\left[1 - \left(\dfrac{v}{c^2}\right)u_x\right]}$$

$$t' = \frac{t}{\left[1 - \left(\dfrac{v}{c^2}\right)\right]}$$

to get the accelerations for mutually attracting frames we should note that the final event will be like that of a frame moving away from the original inertial frame. The original frame can be considered as stationary so that $U_x = 0$ m/s, $U_y = 0$ m/s, $U_z = 0$ m/s and $t = 0$ m/s. The final velocity therefore is V_n which is net of the two velocities and U_x' under Lorentz Velocity Transformations equations. The acceleration will therefore be $a_x = v/t$ which is also the net acceleration of the two accelerations. Under Lorentz Velocity Transformations; $a_x = U_x' /t'$ which reduces to;

$$a'_{xn} = \cfrac{\cfrac{u - V_n}{1 - \left(\frac{v}{c^2}\right)u_x}}{\left[1 - \left(\frac{V_n}{c^2}\right)\right]} \Big/ \cfrac{t}{}$$

$$a'_{xn} = \cfrac{-V_n\left[1 - \left(\frac{V_n}{c^2}\right)\right]}{t}$$

Where a'_{xn} is the net acceleration, along x axis. The negative sign indicates repulsion since the equations are derived from a receding inertial frame. The negative also shows slowing down in our case. When the speed is slow the equations reduces to usual equation of acceleration. When the speed is equal to the speed of light, the acceleration is zero which consistence with the fact that speed of light is constant.

$$a'_{yn} = \cfrac{\cfrac{u_y}{\gamma\left[1 - \left(\frac{V_n}{c^2}\right)u_x\right]}}{\left[1 - \left(\frac{V_n}{c^2}\right)\right]} \Big/ \cfrac{t}{}$$

$$a'_{yn} = \frac{u_y}{\left[1 - \left(\frac{v}{c^2}\right)\right]\gamma}$$

,

$$a'_{zn} = \frac{u_z}{\gamma\left[1 - \left(\frac{V_n}{c^2}\right)u_x\right]} \Bigg/ \frac{t}{\left[1 - \left(\frac{V_n}{c^2}\right)\right]}$$

$$a'_{zn} = \frac{u_z t}{\left[1 - \left(\frac{V_n}{c^2}\right)\right]\gamma}$$

For attracting inertial frames the acceleration

will be;
$$a'_{xn} = \frac{V_n\left[1 - \left(\frac{V_n}{c^2}\right)\right]}{t}$$

$$a'_x = \left| a_n\left[1 - \left(\frac{V_n}{c^2}\right)\right] \right|$$

Where a_n is the net acceleration, $a_n =$
$$\frac{m_1 a_1 + m_2 a_2}{m_1 + m_2}$$

Relativistic Force
For objects attracting we can find the

relativistic net force by considering one frame stationary. So from $F_n = F1 + F2$ which is $F_n = m_1a_1 + m_2a_2$ then fixing one frame gives as its acceleration as zero i.e. $F_n = m_1(0) + m_2a_2$ which reduces to $F_n = m_2a_2$. But a_2 will now be the net acceleration so that we get $F_n = m_2a_n$.

this net acceleration is $a^{nx'}$ under Lorentz transformation. Therefore net relativistic force is $F'_{nx} = m_1 a^{nx'} = \dfrac{mV_n \left[1 - \left(\dfrac{V_{nx}}{c^2} \right) \right]}{t}$

$$F'_{nx} = m_1 a_{nx} \left[1 - \left(\frac{V_{nx}}{c^2} \right) \right]$$

PART 2

A PROPOSED THEORY OF GRAVITY

CHAPTER 1

REDEFINING
ELECTROSTATICS

Present Electrostatic Theory

High school science tells us that there exist two charges; the positive and the negative. The like charges repel and the unlike charges attract.

These are defined arbitrary according to the

object rubbed and by the type of cloth used to rub it by.

The glass rod rubbed with silk is said to be positively charged. The rubber rod rubbed with wool is negatively charged.

The theory to explain this phenomenon is atomic and electron based. It says that the transfer of electrons is the cause of the electrostatic phenomenon. A neutral object has equal numbers of both positive and negative charges so that its overall charge is zero. When rubbed, electrons are rubbed off the object onto the other. So one either has excess or deficit of the electrons. Since the electrons are negatively charged, the object with excess electrons gains a negative charge while the one with a deficit of electrons gains a positive charge, since the protons are positively charged.

Fields

The field for a positive charge was arbitrarily taken to be radial away from the charge or charge carrier, and that of the negative charge radial but toward the charge. This concept is used to explain attraction and repulsion.

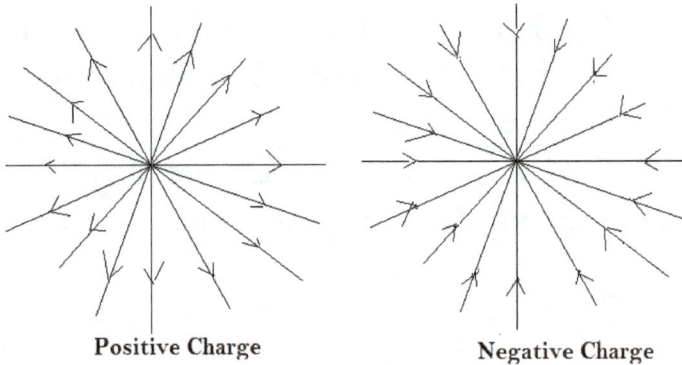

| Positive Charge | Negative Charge |

Definition

A charge is defined as that which gives rise to electricity. Since it gives rise to electricity when it moves, a charge must be considered as a piece of electricity or energy.

Charge must be considered as a qualitative quantity like mass, and not as a particle.

The overall field seen is of the charge carrier not the charges themselves. In short, we don't know the shape of the field. The field is shown as an overall shape of the charge carrier. I therefore propose a 'real' shape of the field of the electron.

Oersted's Discovery

Oersted noticed that a current carrying wire develops a magnetic field around it. This field can be physically observed by sprinkling iron fillings around the wire.

The field was found to be circular all around the wire with the wire as the center.

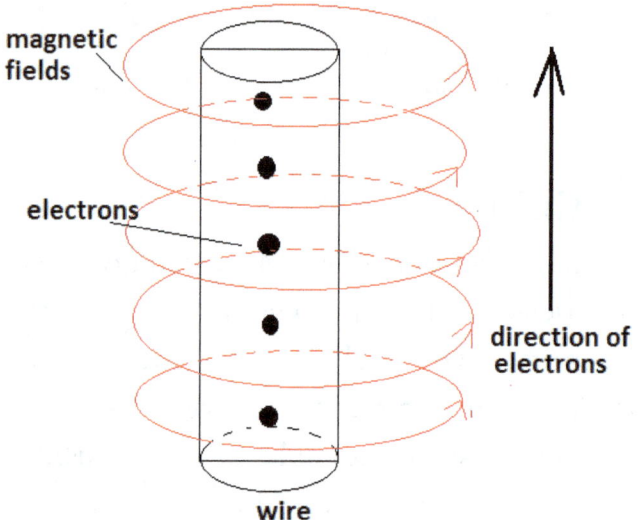

Redefining Charges and Electric Fields

Oersted's phenomenon happens only when the current is flowing. This means that there is a phenomenon which makes the electrons flow and hence create a magnetic field. But the electrons possess a field, a negative field, whether it is moving or not. I propose that the shape of that field around the electron is circular and anticlockwise when the current is moving north as per right hand rule.

This is like taking a slice of the cylindrical electric field around a current carrying wire. When the current is off, the slices of fields point in random directions as per domain theory of magnetism. The potential difference created by a wire connected to a battery makes these slices of fields to align in one direction and fall/move into that field, analogous to an object falling into a gravitational field. The potential difference provides a bigger field which attracts other fields and align them according to the center of mass.

The center of mass is pulled towards the earth in a gravitational field so that a falling object faces the earth or has a face. In the same way an electron has a face. Its center of mass points toward the field. The face is the north pole of the wire or solenoid.

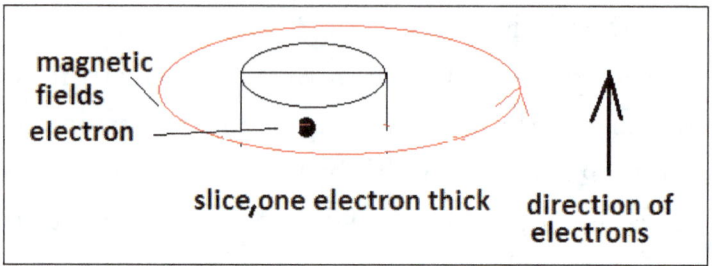

magnetic fields

electron

slice,one electron thick   direction of electrons

An electron has been proved to be a piece of mass so it has a center of gravity/mass. This is the centre of mass which twists in an electric field to face north or in a gravitational field to face down.

When current is off, the center of mass twists to other directions determined by the electrostatic forces of the other particles around it and gravity. Hence the electrons point in random directions. The nucleus scrambles the electrons.

The shape of the fields around a charge or charge carrier has been given as radial for both negative and positive charge.

The 'real' shape of the electric field has not been conclusively observed. Even in the case were it is photographed as radial, it can be

argued that it shows the path of the charged particles not the field itself, just as a piece of nail will be radially attracted to the magnet though the field of the magnet is actually oval and not radial.

Slicing through the magnetic field of a current carrying wire at electron level, we can assume that the field we get is the 'real' field of the electron.

The field for positive or proton can then be said to be clockwise when the proton is moving north. Positive and negative of a charge is now considered wholly in a directional sense and not arbitrary and in terms of which object is rubbed by a particular object.

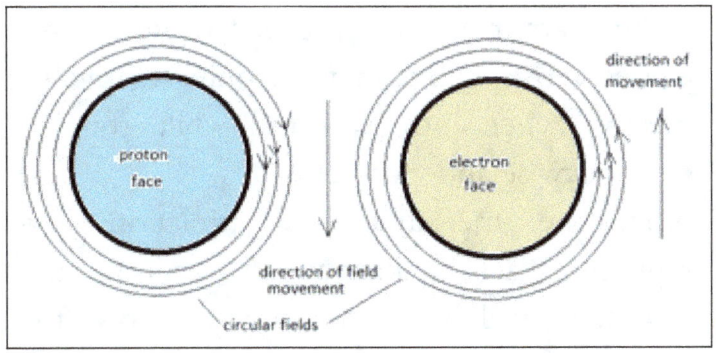

The positive charge carrier is the proton i.e. a piece of mass with a field in clockwise direction when it is facing or moving south. A negative charge carrier is a piece of mass with a field moving anticlockwise when it is moving in the north direction i.e. in opposite direction.

There is only one charge, therefore, which manifests as either positive or negative according to the direction in which the mass is spinning and moving. The definition is based on the direction of the spinning in relation to the direction of the forward movement.

Instead of calling charges positive and negative, we can talk of positive and negative charge carriers. The electron is the negative charge carrier and the proton is the positive charge carrier . But they are only negative or positive when moving north while rotating anticlockwise and vice-versa.

Now the mechanics of attraction and repulsion will change. Attraction and repulsion will be according to how the particles are oriented next to each other. It will be like in two magnets next to each other. In short, the electron and proton effectively become flat circular magnets [pancake magnets].

Ampere's Discovery

Ampere discovered that two current carrying wires with current in the same direction attract each other as the diagram and fields show. The fields merge so the two wires attract.

Two wires with current in the opposite direction repel each other as the diagrams show. The fields collide so the two wires repel.

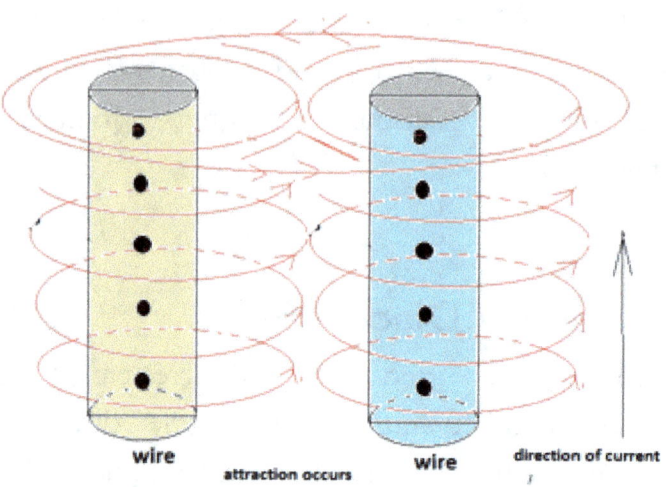

wire   attraction occurs   wire   direction of current

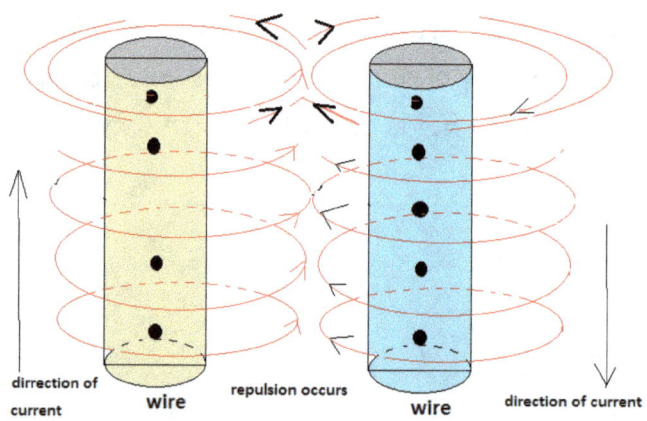

direction of   wire   repulsion occurs   wire   direction of current
current

But oriented face to face i.e. north to north we would expect the wires or solenoids to repel each other. That is; it will be like a solenoid with north-pole facing another north-pole.

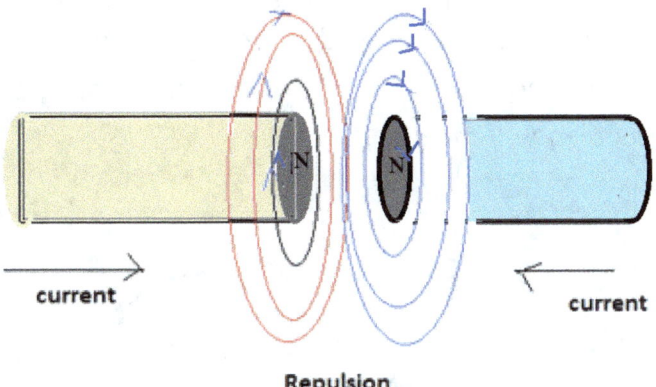

current

current

Repulsion

Sideways Attraction and Repulsion

Ampere 's phenomenon can be used to explain attraction of repulsion between electrons and protons or electrons and electrons. An electron and proton can be either attractive or repulsive next to another electron or proton depending on how they face each other. Electrons in the same direction next to each other will attract sideways. The same goes for protons. Electrons next to each other in opposite directions will repel. The same goes for proton.

Electrons and protons next to each other in the same direction will repel. Electron and proton next to each other in opposite direction will attract.

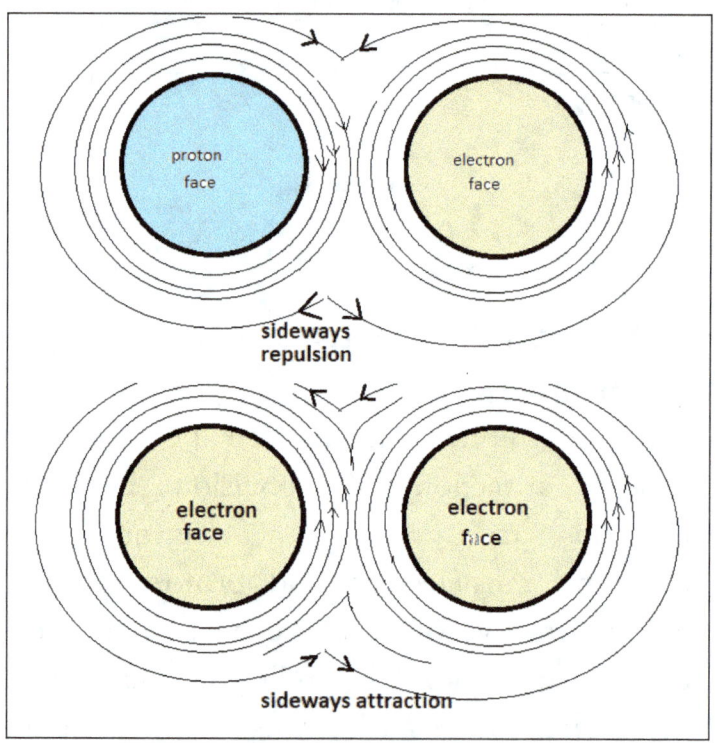

Face to Face Repulsion

If they face each other face to face, they will

repel, and if they face each other face to back, they will attract. The electron will also behave the same.

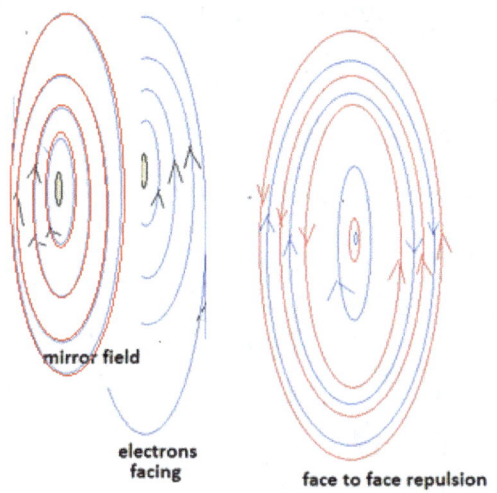

mirror field

electrons
facing

face to face repulsion

 When the electrons face each other face to face, the fields will face in opposite directions. One will be a mirror image of the other.
The two fields therefore collide and repulsion occurs.

Face to Back Attraction

When the electrons are oriented face to back

the fields will be facing each other all in anticlockwise direction i.e. in the same direction.

The fields therefore merge as in magnets of different poles facing each other.

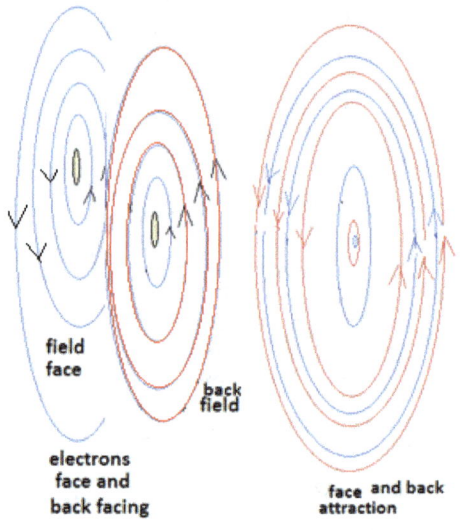

field
face

back
field

electrons
face and
back facing

face and back
attraction

Coulomb's Law

Coulombs law still applies to all the cases because essentially, the field is between electrons and protons and so is electric.

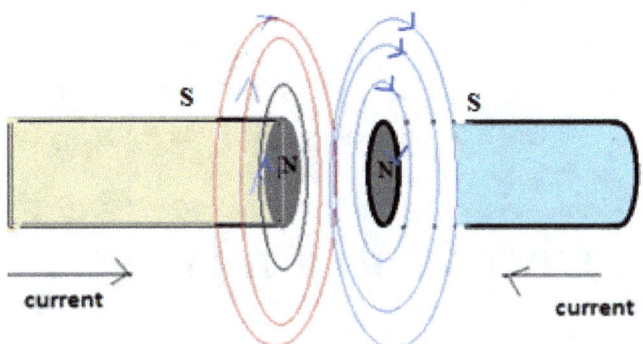

CHAPTER 2

THE CAUSE OF GRAVITY

Polarized Surfaces

The experiment that shows two current carrying wires in the same direction/opposite directions attracting/repelling each other can be applied to the elusive understanding of the cause of gravity.

When the current is on, the wires effectively become magnets and either attract or repel. When the current is off, the wires do not, attract or repel. Nevertheless, the wires can be taken as infinitely many electrons oriented in random directions in the neutral wire i.e. when the current is switched off.

Though the object is neutral, it is slightly negative due to the difference in the positions of the electrons and the protons as per atom theory. The electrons are at the surface and the protons are at the center. In short, all neutral matter is slightly negative or polarized as in polarization of dielectrics in capacitors. The electrons point in random directions so that their fields do not add-up to give a greater field and hence force.

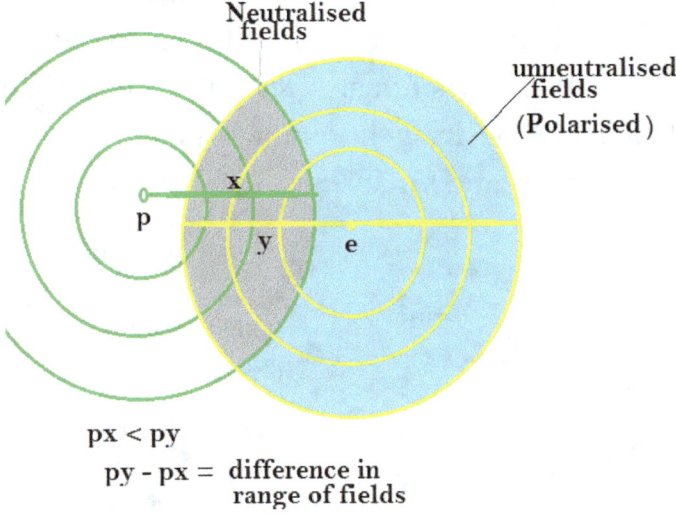

px < py
py - px = difference in range of fields

If x is the radius that positive proton field covers, then the fields are neutralized at the regions were it intersects the fields of the electron up to the end of x, then from x to end of y and beyond the fields are not neutralized i.e. all the space 2y ⁻ x. This region is polarised. It is slightly negative.

Also since the field strength reduces with the factor of $1/r^2$, the field strength at a point P will less for the proton than for the electron because of their difference in position. The electron is nearer to P than the electron though they are in the same atom.

So a neutral surface will be slightly negative. Some of the fields will be pointing up, giving rise to the gravitational force and some of the fields will be lying horizontal, giving rise to the electrostatic force.

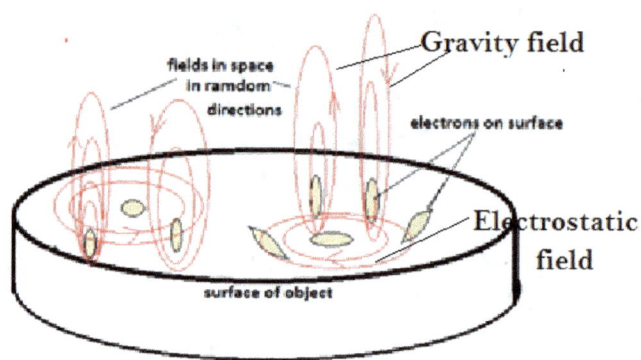

So the two or more neutral objects facing each other are like two wires with current in the same direction. The fields in specific directions will, therefore, hook up just as the two wires attract. The two objects will attract. This is the force we call gravity. It is simply another manifestation of the electric field just as magnetism is a manifestation of the electric field. Other fields will be repelling though. Even if they are vertical and hence gravitational, they will repel if oriented in repulsion position.

But the electrostatic force will also be happening at the same time. The horizontal fields will be attracting or repelling the other horizontal fields on the opposite surface too.

For the fields that are vertical the gravitational force will act upward and the electrostatic force will act sideways. For the fields that are horizontal, the electrostatic force will act upward and the gravitational force will act horizontal. This solves the dilemma of the of how a gravitational field could be induced by an electron when the principle is that electron repels another electron. By redefining positive and negative charge in terms of direction of the field, we can surmount the elusive cause of gravity.

The electric field is therefore the true grand unifying field. The electron is the true boson. All particle physics can now be reduced to just two particles; the electron and the proton. There is now no need to invent a particle that imparts gravity or mass. That makes such a particle a fable. The boson is a fable and unnecessary.

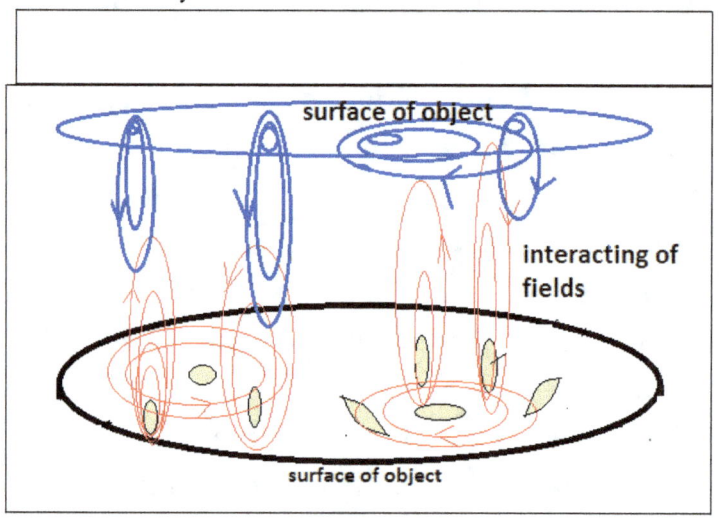

Factor of Polarisation, Я

The surfaces of all neutral objects are slightly charged and the charge is negative. The electron and the proton are in different positions in the atom. This is what polarizes neutral objects. The positive clockwise field of the proton do not align with the negative anticlockwise field of the electron. The electron field overlaps slightly so that there is no total cancellation of the fields. The neutral matter is therefore left slightly negatively charged.

We can introduce a factor to show the amount of Polarization. That factor can be the Coptic symbol Я which can be read as 'backward R'.

The polarising factor, will apply to all masses. It shows the amount of polarisation quantitatively.

Equating the Gravitational Force to Electric Force

The gravitational field at the surface of the earth is given as m $= \dfrac{F_g}{g}$, from F = ma. The electric field due to a source charge, q , is given as E $= \dfrac{F_c}{q}$.

For neutral but polarised matter the electric field can now be given as ;

$$E = \frac{F_c}{Яq}$$

This field is equivalent to the gravitational field at the same point. It can be called the Gravito-Electrical Field and denoted by E_g.

$$\text{so } E_g = \frac{F_c}{Яq}$$

Equating the gravitational field to the gravito-electric field we get;

$$m = E, \quad \frac{F_g}{g} = \frac{F_c}{Яq}, \quad \frac{}{F_g} = \frac{gF_c}{Яq}$$

But Why is Gravity Weak and Electric Force Strong

This force, gravity, is weak because most of it is canceled out by the protons. The difference in the distance between the protons and electrons makes the surface slightly polarized or negative. Because this polarization is small, the gravity is also small.

The electrons on the surface face in random

directions and we would expect some of them to cancel each other out and some to repel. This weakens the force further. However, the protons hold enough electrons in molecules in the same direction more than those that are in random direction. There is therefore a net attractive force between the objects facing each other. There will be enough pairing of them to attract though they maybe in random directions.

Another reason why the electrostatic force maybe greater than gravity is that when electrons face each other face to face or face to back there is more of the field interacting than when they face sideways. There is more surface area when they meet surface to face than sideways. Only the part of the sideways fields where   they meet will interact, and that part is small almost a point.

The static force is greater than gravity because the brushed off electrons orient themselves in one direction with either their faces or bottoms facing up

So the net force is slightly greater than zero. Though some will miss each other, most of them will hook up like in a Velcro zipper.

Redefining Electrostatic Charge of Objects

When rubbed, electrons on the surface of the neutral object are loosed from electrostatic pull of the protons, analogous to wearing off of particles due to friction. These electrons now face up like a magnet's North Pole or South Pole facing up from the surface of the neutral object.

The electrons in the other object have turned upside down so that their bottoms are facing up.

The objects with electrons facing up can be said to be negatively charged. The objects with electrons facing down can be said to be positively charged. Both negativity and positivity being caused by the electron.

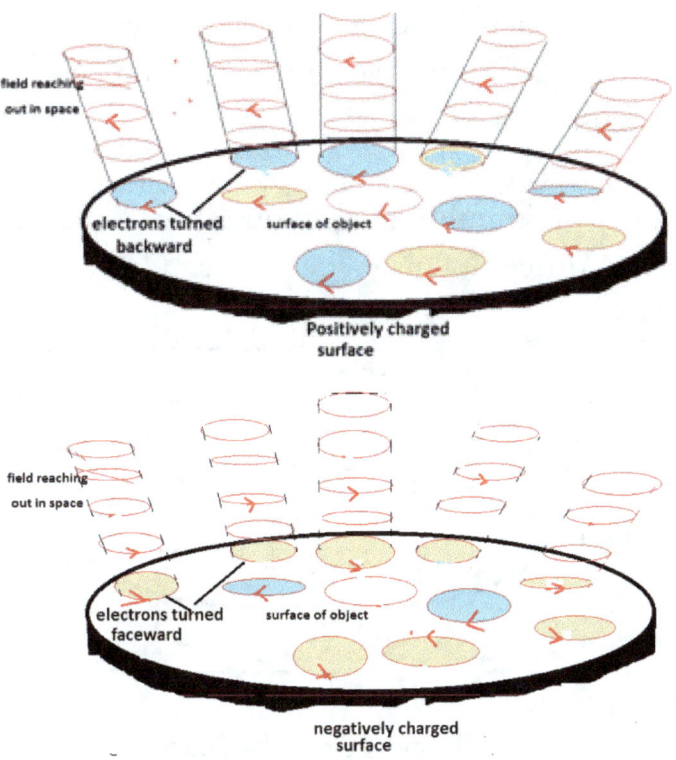

field reaching
out in space

electrons turned
backward

surface of object

**Positively charged
surface**

field reaching
out in space

electrons turned
faceward

surface of object

**negatively charged
surface**

The loosed electrons, under gravity, settle in a stable position either facing up or down or sideways, like a rectangle which can settle in a stable equilibrium either vertical or horizontal under the influence of gravity.

The field on the face of the electron or proton is like that of a North Pole or South Pole of a magnet. Two objects of like charge facing each other is like having two poles of a magnet facing each other. The electrons are facing each other face-ward, so they repel as discussed in chapter 1.

Two differently charged objects will be like two different magnetic poles facing each other.

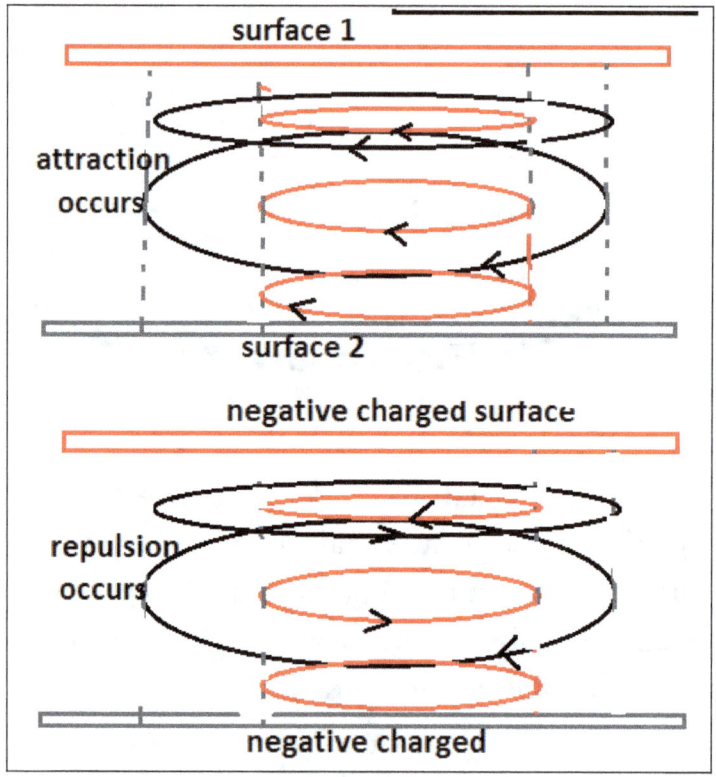

This re-interpretation and theory of electrostatics does away with the jumping of electrons from one object to another, which is more like a chemical reaction, in explaining electrostatic phenomena.

This also helps in explaining positive polarization intuitively. The protons are deep inside the atom so that explaining positive charge on a surface with the idea of electrons being pushed away is problematic. Were exactly will the electrons move inside the atom so that it becomes positive?

Earthing

These loosed electrons are weakly held by the rest of the body so they can flow to the earth or from the earth.

Interaction of Electric and Gravity Field

The charged objects exert gravity and experience gravity because some of the electrons under the loosed electrons are oriented with faces away from the surface so that the rim to rim orientation fields is still happening.

The gravitational field and hence the

gravitational force and the electric field and hence its force interact or interfere with each other just like any other force. They can interact repulsively or attractively, then the net force will be;

$$F_n = \sqrt{F_g{}^2 + F_e{}^2}$$

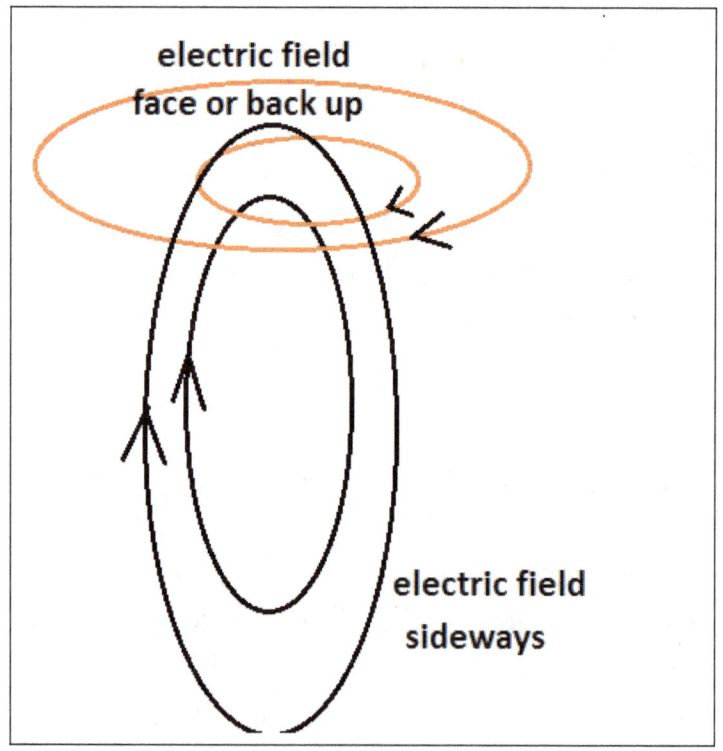

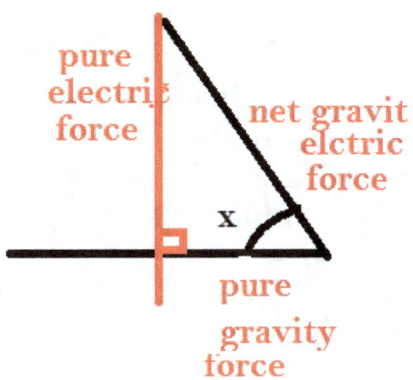

pure electric force

net gravit elctric force

x

pure gravity force

Net gravito-electric force horizontally = gravity force = $F_g{}^2 \cos x$

Net gravito-electric force vertically = electric-force = $F_e{}^2 \sin x$

i.e.
$$F_{ge} = \sqrt{F_g{}^2 \cos x + F_e{}^2 \sin x}$$

i.e.
$$F_{ge} = \sqrt{[G\frac{M_1 M_2}{r^2}]^2 \cos x + [K\frac{Q_1 Q_2}{r^2}]^2 \sin x}$$

But these fields are usually oriented at right angles to each other; it's either the field is vertical or it is horizontal. The Gravito-electrico force, therefore, reduces to gravity force or electrostatic force depending on where the field is pointing i.e. vertical or horizontal on the ground.

$$F_{ge} = \sqrt{[G\frac{M_1M_2}{r^2}]^2 cos90° + [K\frac{Q_1Q_2}{r^2}]^2 sin90°}$$

$$F_{ge} = \sqrt{[G\frac{M_1M_2}{r^2}]^2 [0] + [K\frac{Q_1Q_2}{r^2}]^2 [1]}$$

$$F_{ge} = \sqrt{[K\frac{Q_1Q_2}{r^2}]^2} \quad , \quad F_{ge} = K\frac{Q_1Q_2}{r^2}$$ When the field is horizontal.

OR

$$F_{ge} = \sqrt{[G\frac{M_1M_2}{r^2}]^2 cos90° + [K\frac{Q_1Q_2}{r^2}]^2 sin90°}$$

$$F_{ge} = \sqrt{[G\frac{M_1M_2}{r^2}]^2 [1] + [K\frac{Q_1Q_2}{r^2}]^2 [o]}$$

$$F_{ge} = \sqrt{[G\frac{M_1M_2}{r^2}]^2 [1]} \quad , \quad F_{ge} = G\frac{M_1M_2}{r^2}$$ when the field is vertical.

Equating Coulomb Constant to Gravitational Constant
i) For Electron to Electron
The electrostatic force of the electrons

standing side to side is also the gravitational force since we said it's the source of the gravity. So for electrons facing each other face to face the force is repulsive and negative i.e.

$$F_C = -K \frac{Q_1 Q_2}{r^2} \; .$$

Since it's repulsive, this force cannot be considered as gravitational but electrostatic. Also for vertical fields whose directions collide and hence the force becomes repulsive, the force is not gravitational even if the fields are vertical. Or it can be called repulsive gravitation or negative gravitation.

The attractive force will be considered as the gravitational force and must act sideways. Since it is considered gravitational the polarizing factor for electrons $Я_{ee}$ must be introduced.

So the Coulomb force for fields oriented vertical to vertical will be; $F_{ee} = K \dfrac{Я_{ee} Q_1 Q_2}{r^2}$ instead of $K \dfrac{Q_1 Q_2}{r^2}$. This is the gravitational force.

This polarizing factor in this case can be considered as the result of the difference in distance between when the electrons are face to back and when they are side to side. The greatest attraction occurs face to back because more fields merge than sideways. There is more surface to interact face-to-face than sideways. There are fewer fields sideways than face-ward i.e. the thickness of the field is far much smaller than the width of the field. It is a pancake field or donut field. A slice of the fields one electron thick will merge sideways.

The face-face interaction will make a cylinder implying more fields per area and the side to side will make a plane implying less field per area, for the same distance from the center of the electron.

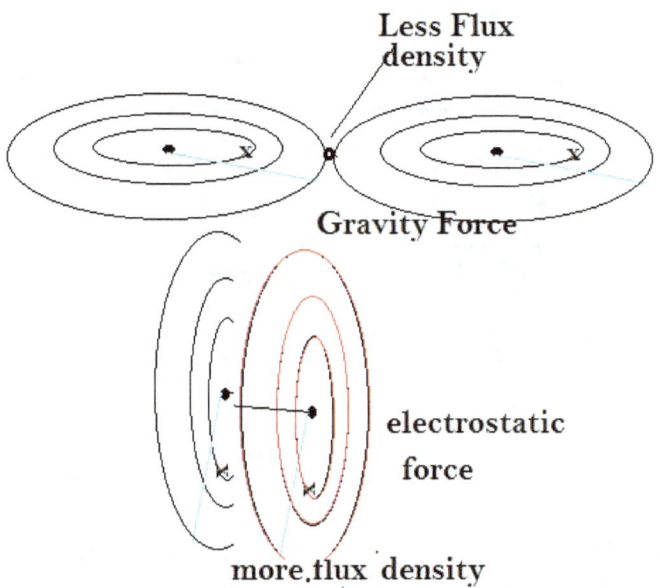

Less Flux density

Gravity Force

electrostatic force

more flux density

So Я$_{ee}$ is determined by the difference in the flux density of the fields when the attraction is sideways and when the attraction is face-to-face i.e. Я$_{ee}$ = ϕ_{E2} − ϕ_{E1}, ϕ_{E2}>ϕ_{E1}.   For proton to proton, it will be Я$_{pp}$ = ϕ_{E2} − ϕ_{E1}. For electron to proton it will be Я$_{ep}$= ϕ_{E2} − ϕ_{E1}.

Equating   K $\dfrac{Я_{ee}Q_1Q_2}{r^2}$ toG $\dfrac{M_1M_2}{r^2}$,   the Gravitational force, we get;

87

$$\frac{Я_{ee}Q_1Q_2}{K\;\;r^2} = G\frac{M_1M_2}{r^2}$$

$$G = K\frac{Я_{ee}Q_1Q_2}{M_eM_e}, \quad G = K\frac{Я_{ee}Q^2}{M^2}$$

$$G = K\,Я_{ee}z^2$$

$$\boxed{G = Я_{ee}Kz^2}$$

But Q/M (z) is charge to mass ratio and was found by J.J. Thomson to be $1.76 \times 10^5 C/kg$.

Magnitude of Я for Particle to Particle

i)   for electron to electron attraction;
$Kz^2 = (1.76 \times 10^8)^2 \, C/kg(8.99 \times 10^9) \, Nm^2/C^2$
$= 27.8 \times 10^{25}$
So G = $Я(27.8 \times 10^{25})$, but G= $6.67 \times 10^{-11}Nm^2/kg^2$
So $Я_{ee} = 6.67 \times 10^{-11}/27.8 \times 10^{25}$,
$Я_{ee} = 0.24 \times 10^{-36} = 2.4 \times 10^{-37}$

For electron to proton $Я_{ep}$ = 4.40 x $10^{-40}N^2m^4/Ckg^4$. This is the $Я_{ep}$ of the hydrogen atom.

This occurs when the two electrons are oriented sideways and pointing in the same direction as the diagrams show.

When the electrons are facing, face to face, [north-north] or back to back, then repulsion will occur. If they are facing face to back attraction will occur. This attraction or repulsion is wholly electrostatic since the fields will fit into each other face to face or face to back leaving no distance between them or their centers.

So now the electron to electron attraction is not wholly repulsive as before, and proton to proton interaction is also not wholly repulsive as before. Both attraction and repulsion will occur according to the orientation of the particles involved.

ii) Electron to Proton

The greatest attraction occurs face to face

with centres of mass touching. A proton to electron Polarising Factor, $Я_{EP}$, must therefore be introduced just as in the first case of the electron to electron attraction.

This polarizing factor in this case can be considered as the result of the difference in distance between when the electrons and proton attraction is face to face and when they are side to side.

So the Coulomb force will be;

$$F_e = -K \frac{Я_{ep}Q_1Q_2}{r^2} \text{, instead of } K = -\frac{Q_1Q_2}{r^2},$$

where r is the distance of the protons from the electrons i.e. radius of the atom.

For a Point P

For an object O at point P the field at that point from the electron will be $E_{eo} = kЯ_{eo}\frac{-Q}{x^2}$,

and from the proton, $E_{po} = kЯ_{po}\frac{Q}{(r+x)^2}$, r = radius of the atom, and x distance from proton to P.

The difference in the fields of the protons and electrons at the surface of the object, i.e. the polarization, will determine the resultant field and hence force experienced by a particle at a point P.

Since the electrons are generally at the surface, the radius of the atom can be taken as the difference in length between the electrons and protons.

So the net field due to difference in length in that direction at that point will be;

$$E_n = E_p + E_e$$

i.e. $E_n = k\text{Я}_{po}\dfrac{Q}{(r+x)^2} + k\text{Я}_{eo}\dfrac{-Q}{x^2}$

$E_n = kQ(\dfrac{\text{Я}_{po}}{(r+x)^2} - \dfrac{\text{Я}_{eo}}{x^2})$, this electric field in this orientation is now called the Gravity Field, E_g.

So $E_g = kQ(\dfrac{\text{Я}_{po}}{(r+x)^2} - \dfrac{\text{Я}_{eo}}{x^2})$

So the electric force experienced by a test charge q /mass O at P in that direction is ;

$$F_e = E_g q = \frac{kQq(\frac{\text{Я}_{po}}{(r+x)^2} - \frac{\text{Я}_{eo}}{x^2})}{}$$, but this is the pure gravitational force because of the orientation.

$$\boxed{\text{So } F_g = E_g q =} \; kQq(\frac{\text{Я}_{po}}{(r+x)^2} - \frac{\text{Я}_{eo}}{x^2})$$

The Gravitational Force at P Near the Ground

This electric field on the test mass/charge at this point and near the ground from a single electron is equal to the gravitational field at this point.

That is; $$kQq(\frac{\text{Я}_{po}}{(r+x)^2} - \frac{\text{Я}_{eo}}{x^2}) = \frac{F_g}{g}$$

So $F_g = kQqg\left(\frac{\text{Я}_{po}}{(r+x)^2} - \frac{\text{Я}_{eo}}{x^2}\right)$, This is the Gravity Force experienced by a charge/object O near the ground because of g.

If many electrons are acting on P then

$F_g = kNQqg\left(\frac{\text{Я}_{po}}{(r+x)^2} - \frac{\text{Я}_{eo}}{x^2}\right)$, N being the number of electrons attracting O at P.

For two objects near each other with gravitational fields g_1, g_2 then the net Force will be;

$$F_g = 2kNQq\left(\frac{Я_{po}}{(r+x)^2} - \frac{Я_{eo}}{x^2}\right)(g_e + g_o)$$

Or $F_g = 2kNQq\left(\frac{Я_{po}}{(r+x)^2} - \frac{Я_{eo}}{x^2}\right)g_n$, $g_n = g_e + g_o$, i.e. $g_n = $ net $g = g$.

But two masses or charges will also be interacting pure-electrostatically repulsively or attractively. This electrostatic force is given by the classic formula; $F_e = K\dfrac{Qq}{r^2}$ if attractive

or $\qquad\qquad F_e = \dfrac{-K\dfrac{Qq}{r^2}}{}$ if repulsive.

So the net gravito-electrico force, F_{ge}, near the surface of the earth is;

$$F_{ge} = \sqrt{F_g{}^2 cosx + F_e{}^2 sinx}$$

$$F_{ge} = \sqrt{[2kNQqg\left(\frac{Я_{po}}{(r+x)^2} - \frac{Я_{eo}}{x^2}\right)]^2 cosx + [K\frac{NQq}{x^2}]^2 sinx}$$

This is the force an object near a planet or earth experiences. It is not purely gravity, but the net of gravity and electrostatic forces.

iii) Polarising Factor for Neutral Objects (Equating G to K)

For two neutral objects of masses M_1 and M_2 separated by a distance of x meters. The two objects can be considered to be charged objects with charges Q and q. The Gravito-Coulomb force can therefore apply to the objects.

The force of gravity between them is given as;

$$F_g = G\,\frac{M_1 M_2}{x^2}$$

Equating the two we get; $F_c = F_g$

$$kQq\left(\frac{Я_{po}}{(r+x)^2} - \frac{Я_{eo}}{x^2}\right) = G\,\frac{M_1 M_2}{x^2}$$

$$G = \frac{KQq}{M_1 M_2}\left(\frac{Я_{po}}{(r+x)^2} - \frac{Я_{eo}}{x^2}\right)x^2$$

$$\boxed{G = \frac{KQq}{M_1 M_2}\left(\frac{Я_{po}x^2}{(r+x)^2} - Я_{eo}\right)}$$

For the whole object the total polarised charge will be number of electrons oriented for attraction multiplied by the charge.

$$\text{So} \quad G = \frac{KNQq}{M_1 M_2}\left(\frac{Я_{po}x^2}{(r+x)^2} - Я_{eo}\right)$$

N is number of pairs of electrons oriented for attraction.

Not all electrons will be oriented in such a way that attraction occurs. Some will repel each other and others will be oriented perpendicular to the sideways field so that not attractive nor repulsive force will occur. So it is possible to have more than half of the surface electrons not oriented in attractive way but still have a net attractive force.

But Q/M, $Ƶ$, [z cross] can be called the polarised charge to mass electron density or charge to mass ratio.

$$\text{So} \quad G = KN Ƶ_1 Ƶ_2 \left(\frac{Я_{po}x^2}{(r+x)^2} - Я_{eo}\right)$$

$\frac{e}{m}$ is equal to the charge to mass ratio. But because N is a fraction of the total number of charges, the overall charge is smaller compared to the charge to mass ratio of the electron to electron attraction. This will reduce the Coulomb force in comparison to the electron to electron force.

The value $N\left(\frac{Я_{po}x^2}{(r+x)^2} - Я_{eo}\right)$ can be called the Polarisation factor of neutral objects, Я.

So $$Я = N\left(\frac{Я_{po}x^2}{(r+x)^2} - Я_{eo}\right) \text{ and}$$

$$\boxed{G = KЯz_1}z_2$$

The Net Gravito-Electrico Field between Two Neutral Objects

For two neutral objects of masses M1 and M2 separated by a distance x with N pairs of electron to electron attraction, the net field will be

$E = E_2 - E_1,$

$$E = kQN\left(\frac{Я_{po}}{(r+x)^2} - \frac{Я_{eo}}{x^2}\right) - \left[-kQN\left(\frac{Я_{po}}{(r+x)^2} - \frac{Я_{eo}}{x^2}\right)\right]$$

$$E_g = 2\ kQN\left(\frac{Я_{po}}{(r+x)^2} - \frac{Я_{eo}}{x^2}\right) \quad \text{and net force}$$

$$F_g = 2kNQq\left(\frac{Я_{po}}{(r+x)^2} - \frac{Я_{eo}}{x^2}\right) = 2kЯQq$$

$$\text{Or} \quad F_g = \frac{2G}{Яz_1z_2}NQq\left(\frac{Я_{po}}{(r+x)^2} - \frac{Я_{eo}}{x^2}\right) = \frac{2G}{z_1z_2}Qq$$

Then there is the pure electrostatic forces acting on the masses too.

The net gravito-electrico force is,

$$F_{ge} = \sqrt{\left[2kNQq\left(\frac{Я_{po}}{(r+x)^2} - \frac{Я_{eo}}{x^2}\right)\right]^2 cosx + \left[K\frac{NQq}{x^2}\right]^2 sinx}$$

OR

$$F_{ge} = \sqrt{\left[2kЯQq\right]^2 cosx + \left[K\frac{NQq}{x^2}\right]^2 sinx}$$

OR

$$F_{ge} = \sqrt{\left[\frac{2G}{Яz_1z_2}NQq\left(\frac{Я_{po}}{(r+x)^2} - \frac{Я_{eo}}{x^2}\right)\right]^2 cosx + \left[K\frac{NQq}{x^2}\right]^2 sinx}$$

OR

$$F_{ge} = \sqrt{\left[\frac{2G}{z_1z_2}Qq\right]^2 cosx + \left[K\frac{NQq}{x^2}\right]^2 sinx}$$

This is the gravity experienced by masses far astronomically apart. This is the Universal Gravito-electrico Force.

REFERENCES

1. Nelkon M, 1993, Principles of Physics, Longman, London.
2. Serway, R. and Jewett, J. 2004, 6th Ed, Physics for Scientists and Engineers, Thomas Brooks/Cole,
3. Thornton, S. Rex, A. 2013 4th ed. Modern Physics for Scientists and Engineers, Brooks/Cole, Boston.
4. Zumdahl, S. 2003, Chemical Principles, 5th ed. Houghton Mifflin Company
5. Demana F, Waits B, Clemens S, 1994, 3RD Edition, Precalculus Mathematics; Addison –Wesley,Californa.
6. Various Internet Resources